YEAR 1

D1147121

STAR MATHS STARTERS

A fresh approach to mental maths

Please check disc at front on issue and discharge

TERMS AND CONDITIONS

IMPORTANT – PERMITTED USE AND WARNINGS – READ CAREFULLY BEFORE USING

Minimum specification:
- PC with a CD-ROM drive and 512 Mb RAM (recommended)
- Windows 98SE or above/Mac OSX.1 or above
- Recommended minimum processor speed: 1 GHz
- Facilities for printing

Julie Cogill and Anthony David

3.2011

Authors
Julie Cogill and Anthony David

Anthony David dedicates this book to his wife Peachey, and sons Oliver and Samuel.

Editors
Niamh O'Carroll and Roanne Charles

Assistant Editors
Pollyanna Poulter and Margaret Eaton

Illustrator
Theresa Tibbetts (Beehive Illustration)

Series Designer
Joy Monkhouse

Designers
Allison Parry and Melissa Leeke

Text © Julie Cogill and Anthony David
© 2008 Scholastic Ltd

CD-ROM development in association with Vivid Interactive

Designed using Adobe InDesign and Adobe Illustrator

Published by Scholastic Ltd
Book End, Range Road, Witney,
Oxfordshire OX29 0YD
www.scholastic.co.uk

Printed by Bell & Bain, Glasgow
2 3 4 5 6 7 8 9   0 1 2 3 4 5 6 7

ISBN 978-1407-10007-4

ACKNOWLEDGEMENTS
Extracts from the Primary National Strategy's *Primary Framework for Mathematics* (2006)
www.standards.dfes.gov.uk/primaryframework, *Renewing the Primary Framework* (2006) and the
Interactive Teaching Programs originally developed for the National Numeracy Strategy © Crown
copyright. Reproduced under the terms of the Click Use Licence.

Every effort has been made to trace copyright holders for the works reproduced in this book, and the
publishers apologise for any inadvertent omissions.

British Library Cataloguing-in-Publication Data
A catalogue record for this book is available from the British Library.

Mixed Sources
Product group from well-managed
forests and other controlled sources
www.fsc.org Cert no. TT-COC-002769
© 1996 Forest Stewardship Council
FSC

Introduction

In the 1999 *Framework for Teaching Mathematics* the first part of the daily mathematics lesson is described as 'whole-class work to rehearse, sharpen and develop mental and oral skills'. The Framework identified a number of short, focused activities that might form part of this oral and mental work. Teachers responded very positively to these 'starters' and they were often judged by Ofsted to be the strongest part of mathematics lessons.

However, the renewed *Primary Framework for Mathematics* (2006) highlights that the initial focus of 'starters', as rehearsing mental and oral skills, has expanded to become a vehicle for teaching a range of mathematics. 'Too often the "starter" has become an activity extended beyond the recommended five to ten minutes' (*Renewing the Primary Framework for mathematics: Guidance paper,* 2006). The renewed Framework also suggests that 'the focus on oral and mental calculation has been lost and needs to be reinvigorated'.

Star Maths Starters aims to 'freshen up' the oral and mental starter by providing focused activities that help to secure children's knowledge and sharpen their oral and mental skills. It is a new series, designed to provide classes and teachers with a bank of stimulating interactive whiteboard resources for use as starter activities. Each of the 30 starters offers a short, focused activity designed for the first five to ten minutes of the daily mathematics lesson. Equally, the starters can be used as stand-alone oral and mental maths 'games' to get the most from a spare ten minutes in the day.

About the book

Each book includes a bank of teachers' notes linked to the interactive whole-class activities on the CD-ROM. A range of additional support is also provided, including planning grids, classroom resources, generic support for using the interactive whiteboard in mathematics lessons, and an objectives grid.

Objectives grid

A comprehensive two-page planning grid identifies links to the *Primary Framework for Mathematics* strands and objectives. The grid also identifies one of six starter types, appropriate to each interactive activity (see page 7 for further information).

Starter Number	Star Starter Title	Page No.	Strand	Learning objective as taken from the Primary Framework for Mathematics	Type of Starter
1	Targets: counting or adding	13	Using and applying mathematics	Solve problems involving counting or adding	Rehearse
2	Making patterns	14	Using and applying mathematics	Describe simple patterns and relationships involving numbers or shapes; decide whether examples satisfy given conditions	Read
3	Twenty cards (ITP): ordering	15	Counting and understanding number	Compare and order numbers, using the related vocabulary	Refine
4	Bricks: numbers to 20	16	Counting and understanding number	Compare and order numbers, using the related vocabulary	Rehearse Reason
5	Dominoes: numbers and names	17	Counting and understanding number	Read and write numerals from 0 to 20	Refresh
6	Number line: 1 more, 1 less	18	Counting and understanding number	Say the number that is 1 more or less than any given number up to 10	Recall
7	Counting on and back (ITP)	19	Counting and understanding number	Say the number that is 1 more or less than any given number, and 10 more or less for multiples of 10	Refine
8	Bricks: multiples of 10	20	Counting and understanding number	Say the number that is 10 more or less for multiples of 10	Rehearse Reason

Highlighted text indicates the end-of-year objectives

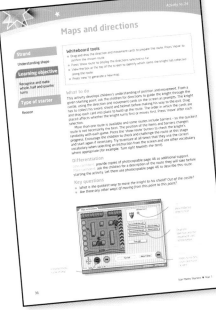

Activity pages

Each page of teachers' notes includes:

Learning objectives

Covering the strands and objectives of the renewed *Primary Framework for Mathematics*

Type of starter

Identifying one or more of the 'six Rs' of oral and mental work (see page 7)

Whiteboard tools

Identifying the key functions of the accompanying CD-ROM activity

What to do

Outline notes on how to administer the activity with the whole class

Differentiation

Adapting the activity for more or less confident learners

Key questions

Probing questions to stimulate and sustain the oral and mental work

Annotations

At-a-glance instructions for using the CD-ROM activity.

Whiteboard hints and tips

Each title offers some general support identifying practical mathematical activities that can be performed on any interactive whiteboard (see pages 8–9).

Recording sheets

Two recording sheets have been included to support your planning:

● Planning for the 'six Rs': plan a balance of activities across the six Rs of mental and oral maths (see page 7).
● Star Maths Starters diary: build a record of the starters used (titles, objectives covered, how they were used and dates they were used).

About the CD-ROM

Types of activity

Each CD-ROM contains 30 interactive starter activities for use on any interactive whiteboard. These include:

Interactive whiteboard resources

A set of engaging interactive activities specifically designed for *Star Maths Starters*. The teachers' notes on pages 13–42 of this book explain how each activity can be used for a ten-minute mental maths starter, with annotated screen shots giving you at-a-glance support. Similarly, a 'what to do' function within each activity provides at-the-board support.

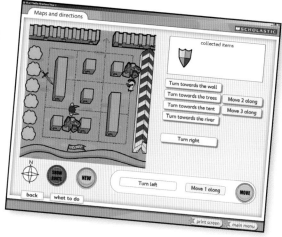

Interactive teaching programs (ITPs)

A small number of ITPs, originally developed by the National Numeracy Strategy, has been included on each CD-ROM. They are simple programs that model a range of objectives, such as data presentation or fraction bars. Their strength is that they are easy to read and use. If you press the Esc button the ITP will reduce to a window on the computer screen. It can then be enlarged or more ITPs can be launched and set up to model further objectives, or simply to extend the objective from that starter. To view the relevant 'what to do' notes once an ITP is open, press the Esc button to gain access to the function on the opening screen of the activity.

Interactive 'notepad'

A pop-up 'notepad' is built into a variety of activities. This allows the user to write answers or keep a record of workings out and includes 'pen', 'eraser' and 'clear' tools.

Teacher zone

This teachers' section includes links from the interactive activities to the *Primary Framework for Mathematics* strands, together with editable objectives grids, planning grids and printable versions of the activity sheets on pages 43–46.

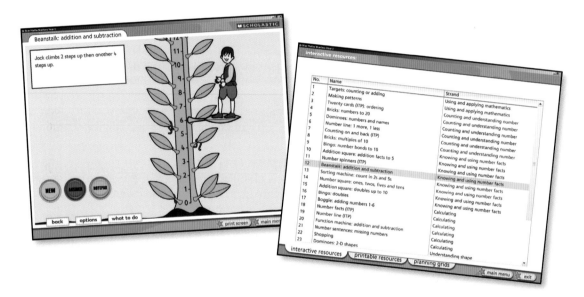

How to use the CD-ROM

System requirements

Minimum specification
- PC with a CD-ROM drive and 512 Mb RAM (recommended)
- Windows 98SE or above/Mac OSX.1 or above
- Recommended minimum processor speed: 1 GHz

Getting started

The *Star Maths Starters* CD-ROM should auto run when inserted into your CD drive. If it does not, use **My Computer** to browse the contents of the CD-ROM and click on the 'Star Maths Starters' icon.

From the start-up screen you will find four options: select **Credits** to view a list of credits. Click on **Register** to register the product to receive product updates and special offers. Click on **How to use** to access support notes for using the CD-ROM. Finally, if you agree to the terms and conditions, select **Start** to move to the main menu.

For all technical support queries, please phone Scholastic Customer Services on 0845 6039091.

The six Rs of oral and mental work

In the guidance paper *Renewing the Primary Framework for mathematics* (2006), the Primary National Strategy identified six features of children's mathematical learning that oral and mental work can support. The description of the learning and an outline of possible activities are given below:

Six Rs	Learning focus	Possible activities
Rehearse	To practise and consolidate existing skills, usually mental calculation skills, set in a context to involve children in problem solving through the use and application of these skills; use of vocabulary and language of number, properties of shapes or describing and reasoning.	Interpret words such as more, less, sum, altogether, difference, subtract; find missing numbers or missing angles on a straight line; say the number of days in four weeks or the number of 5p coins that make up 35p; describe part-revealed shapes, hidden solids; describe patterns or relationships; explain decisions or why something meets criteria.
Recall	To secure knowledge of facts, usually number facts; build up speed and accuracy; recall quickly names and properties of shapes, units of measure or types of charts, graphs to represent data.	Count on and back in steps of constant size; recite the 6-times table and derive associated division facts; name a shape with five sides or a solid with five flat faces; list properties of cuboids; state units of time and their relationships.
Refresh	To draw on and revisit previous learning; to assess, review and strengthen children's previously acquired knowledge and skills relevant to later learning; return to aspects of mathematics with which the children have had difficulty; draw out key points from learning.	Refresh multiplication facts or properties of shapes and associated vocabulary; find factor pairs for given multiples; return to earlier work on identifying fractional parts of given shapes; locate shapes in a grid as preparation for lesson on coordinates; refer to general cases and identify new cases.
Refine	To sharpen methods and procedures; explain strategies and solutions; extend ideas and develop and deepen the children's knowledge; reinforce their understanding of key concepts; build on earlier learning so that strategies and techniques become more efficient and precise.	Find differences between two two-digit numbers, extend to three-digit numbers to develop skill; find 10% of quantities, then 5% and 20% by halving and doubling; use audible and quiet counting techniques to extend skills; give coordinates of shapes in different orientations to hone concept; review informal calculation strategies.
Read	To use mathematical vocabulary and interpret images, diagrams and symbols correctly; read number sentences and provide equivalents; describe and explain diagrams and features involving scales, tables or graphs; identify shapes from a list of their properties; read and interpret word problems and puzzles; create their own problems and lines of enquiry.	Tell a story using an interactive bar chart; alter the chart for children to retell the story; starting with a number sentence (eg 2 + 11 = 13), children generate and read equivalent statements for 13; read values on scales with different intervals; read information about a shape and eliminate possible shapes; set number sentences in given contexts; read others' results and offer new questions and ideas for enquiry.
Reason	To use and apply acquired knowledge, skills and understanding; make informed choices and decisions, predict and hypothesise; use deductive reasoning to eliminate or conclude; provide examples that satisfy a condition always, sometimes or never and say why.	Sort shapes into groups and give reasons for selection; discuss why alternative methods of calculation work and when to use them; decide what calculation to do in a problem and explain the choice; deduce a solid from a 2D picture; use fractions to express proportions; draw conclusions from given statements to solve puzzles.

Each one of the styles of starter enables children to access different mathematical skills and each has a different outcome, as identified above. A bingo game, for example, provides a good way of rehearsing number facts, whereas a 'scales' activity supports reading skills. In the objectives grid on pages 10–11, the type of each Star Maths Starters activity is identified to make it easier to choose appropriate styles of starter matched to a particular objective. A 'Six Rs' recording sheet has also been provided on page 12 (with an editable version on the CD-ROM) to track the types of starter you will be using against the strands of the renewed Framework.

Using the interactive whiteboard in primary mathematics

The interactive whiteboard is an invaluable tool for teaching and learning mathematics. It can be used to demonstrate and model mathematical concepts to the whole class, offering the potential to share children's learning experiences. It gives access to powerful resources - audio, video, images, websites and interactive activities - to discuss, interact with and learn from. *Star Maths Starters* provides 30 quality interactive resources that are easy to set up and use and which help children to improve their mathematical development and thinking skills through their use as short, focused oral and mental starters.

Whiteboard resources and children's learning

There are many reasons why the whiteboard, especially in mathematics, enhances children's learning:

- Using high-quality interactive maths resources will engage children in the process of learning and developing their mathematical thinking skills. Resources such as maths games can create a real sense of theatre in the whole class and promote a real desire to achieve and succeed in a task.
- As mentioned above, the whiteboard can be used to demonstrate some very important mathematical concepts. For example, many teachers find that children understand place value much faster and more thoroughly through using interactive resources on a whiteboard. Similarly, the whiteboard can support children's visualisation of mathematics, especially for 'Shape and Space' activities.
- Although mathematics usually has a correct or incorrect answer, there are often several ways of reaching the same result. The whiteboard allows the teacher to demonstrate methods and encourages children to present and compare their own mental or written methods of calculation.

Using a whiteboard in Year 1

An interactive whiteboard can be used for a variety of purposes in Year 1 mathematics lessons. These include:

- using clipart to demonstrate simple addition and subtraction (for example, moving sheep in and out of a pen);
- using interactive 100-squares to highlight multiples and show number patterns (for example, showing the result of doubling and halving numbers);
- estimating and counting prepared sets of clipart objects - counting from a distance adds an extra degree of challenge to the children, so use the whiteboard tools to highlight sets or patterns to support them as appropriate;
- using clipart or drawing tools to present pictures and patterns using 2D shapes;
- using software programs to collect and organise data and show the results of class surveys in simple block graphs.

Practical considerations

For the teacher, the whiteboard has the potential to save preparation and classroom time, as well as providing more flexible teaching.

ICT resources for the interactive whiteboard often involve numbers that are randomly generated, so that possible questions or calculations stemming from a single resource may be many and varied. This enables resources to be used for a longer or shorter time period depending on the purpose of the activity and how children's learning is progressing. *Star Maths Starters* includes many activities of this type.

From the very practical point of view of saving teachers' time, particularly in the starter activity, it is often easier to set up mathematics resources more quickly than those for other subjects. Once the software is familiar, preparation time is saved especially when there is need for clear presentation, as in drawing shapes accurately or creating charts and diagrams for 'Handling data' activities.

Maths resources on the interactive whiteboard are often flexible and enable differentiation so that a teacher can access different degrees of difficulty using the same software. Last but not least, whiteboard resources save time writing on the board and software often checks calculations, if required, which enables more time both for teaching and assessing children's understanding.

Using *Star Maths Starters* interactively

Much has been said and written about interactivity in the classroom but it is not always clear what this means. For example, children coming out to the board and ticking a box is not what is meant by 'whole-class interactive teaching and learning'. In mathematics it is about challenging children's ideas so that they develop their own thinking skills and, when appropriate, encouraging them to make connections across different mathematical topics. As a teacher, this means asking suitable questions and encouraging children to explore and discuss their methods of calculation and whether there are alternative ways of achieving the same result. *Star Maths Starters* provides some examples of key questions that could be asked while the activities are being undertaken, together with suggestions for how to engage less confident learners and stretch the more confident.

If you already have some experience in using the whiteboard interactively then we hope the teaching suggestions set out in this book will take you further. What is especially important is the facility the whiteboard provides to share pupils' mathematical learning experiences. This does not mean just asking children to suggest answers, but using the facility of the board to display and discuss ideas so that everyone can share in the learning experience. Obviously, this needs to be in a way that explores and relates the thinking of individuals to the context of the learning that is happening.

In the best whiteboard classrooms, teachers comment that the board provides a shared learning experience between the teacher and the class, in so far as the teacher may sometimes stand aside while children themselves are discussing their own mathematical methods and ideas.

Starter Number	Star Starter Title	Page No.	Strand	Learning objective as taken from the Primary Framework for Mathematics	Type of Starter
1	Targets: counting or adding	13	Using and applying mathematics	Solve problems involving counting or adding	Rehearse
2	Making patterns	14	Using and applying mathematics	Describe simple patterns and relationships involving numbers or shapes; decide whether examples satisfy given conditions	Read
3	Twenty cards (ITP): ordering	15	Counting and understanding number	Compare and order numbers, using the related vocabulary	Refine
4	Bricks: numbers to 20	16	Counting and understanding number	Compare and order numbers, using the related vocabulary	Rehearse Reason
5	Dominoes: numbers and names	17	Counting and understanding number	Read and write numerals from 0 to 20	Refresh
6	Number line: 1 more, 1 less	18	Counting and understanding number	Say the number that is 1 more or less than any given number up to 10	Recall
7	Counting on and back (ITP)	19	Counting and understanding number	Say the number that is 1 more or less than any given number, and 10 more or less for multiples of 10	Refine
8	Bricks: multiples of 10	20	Counting and understanding number	Say the number that is 10 more or less for multiples of 10	Rehearse Reason
9	Bingo: number bonds to 10	21	Knowing and using number facts	Derive and recall all pairs of numbers with a total of 10	Rehearse
10	Addition square: addition facts to 5	22	Knowing and using number facts	Recall addition facts for totals to at least 5	Recall
11	Number spinners (ITP)	23	Knowing and using number facts	Derive and recall all addition facts for totals to at least 5; work out the corresponding subtraction facts	Refine
12	Beanstalk: addition and subtraction	24	Knowing and using number facts	Derive addition facts for totals to at least 5; work out the corresponding subtraction facts	Recall
13	Sorting machine: count in twos and fives	25	Knowing and using number facts	Count on in twos and fives and use this knowledge to derive the multiples of 2 and 5	Reason
14	Number square: ones, twos, fives and tens	26	Knowing and using number facts	Count on and back in ones, twos, fives and tens and use this knowledge to derive the multiples of 2, 5 and 10	Recall
15	Addition square: doubles to 10	27	Knowing and using number facts	Recall the doubles of all numbers to at least 10	Recall

Starter Number	Star Starter Title	Page No.	Strand	Learning objective as taken from the Primary Framework for Mathematics	Type of Starter
16	Bingo: doubles	28	Knowing and using number facts	Recall the doubles of all numbers to at least 10	Rehearse
17	Maths Boggle: adding numbers 1 to 6	29	Calculating	Recognise that addition can be done in any order	Rehearse
18	Number facts (ITP)	30	Calculating	Understand subtraction as 'take away' and find a 'difference' by counting up	Refine
19	Number line (ITP)	31	Calculating	Find a 'difference' by counting up	Refine
20	Function machine: addition and subtraction	32	Calculating	Use the vocabulary related to addition and subtraction to describe number sentences	Read Reason
21	Number sentences: missing numbers	33	Calculating	Use the vocabulary related to addition and subtraction and symbols to describe and record addition and subtraction number sentences	Read
22	Shopping: money problems	34	Calculating	Solve practical problems that involve combining groups of 2, 5 or 10	Refine
23	Dominoes: 2D shapes	35	Understanding shape	Visualise and name common 2D shapes	Refresh Read
24	Maps and directions	36	Understanding shape	Recognise and make whole, half and quarter turns	Reason
25	Find that cat!	37	Understanding shape	Visualise and use everyday language to describe the position of objects	Reason
26	Pan balance: comparing weights	38	Measuring	Estimate, weigh and compare objects	Rehearse
27	Measuring jug	39	Measuring	Use suitable non-standard or standard units and measuring instruments (eg a measuring jug)	Refine
28	Clock	40	Measuring	Use vocabulary related to time; read the time to the hour and half hour	Read
29	Pictogram: favourite foods	41	Handling data	Answer a question by recording information; present outcomes using pictograms	Refine
30	Block graph: favourite pets	42	Handling data	Answer a question by recording information; present outcomes using block graphs	Refine

Planning for the six Rs of oral and mental work

Oral and mental activity – six Rs	Using and applying mathematics	Counting and understanding number	Knowing and using number facts	Calculating	Understanding shape	Measuring	Handling data
Rehearse	• Targets: counting or adding	• Bricks: numbers to 20 • Bricks: multiples of 10	• Bingo: number bonds to 10 • Bingo doubles	• Maths Boggle: adding numbers 1 to 6		• Pan balance: comparing weights	
Recall		• Number line: 1 more, 1 less	• Addition square: addition facts to 5 • Beanstalk: addition and subtraction • Number square: ones, twos, fives and tens • Addition square: doubles to 10				
Refresh		• Dominoes: numbers and names			• Dominoes: 2D shapes		
Refine		• Twenty cards (ITP): ordering • Counting on and back (ITP)	• Number spinners	• Number facts (ITP) • Number line (ITP) • Shopping: money problems		• Measuring jug	• Pictogram: favourite foods • Block graph: favourite pets
Read	• Number square: ones, twos, fives and tens			• Function machine: addition and subtraction • Numbers sentences: missing numbers	• Dominoes: 2D shapes	• Clock	
Reason		• Bricks: numbers to 20 • Bricks: multiples of 10	• Sorting machine: count in twos and fives	• Function machine: addition and subtraction	• Maps and directions • Find that cat!		

Targets: counting or adding

Strand

Using and applying mathematics

Learning objective

Solve problems involving counting or adding

Type of starter

Rehearse

Whiteboard tools

- Press 'go' to deal three number cards.
- Press the blue target circle to generate a target number.
- Use the 'pen' tool in the 'notepad' to show calculations. Press 'start again' to delete any text from the notepad, or use the 'eraser' tool.
- Press 'winner' if the children complete the activity successfully.

What to do

The aim of this activity is for the children to use counting or addition to find a target number. Each target number should be achievable by counting on from, or adding, the numbers shown on the three cards. Press the blue target circle to generate a target number or, alternatively, write a target number (using the pen tool in the on-screen notepad) into the target circle. Encourage the children to count on from one card to the next, or to add two or more of the cards, to match (or nearly match) the target number. Invite the children to write their answers on their individual whiteboards and discuss how the answers were reached. Help them to explain their calculations, using the pen tool in the notepad to help, and review the process that each child has followed to reach each target number.

Differentiation

Less confident: provide counting blocks (or similar) as a visual cue and ask another adult to support the children with their calculations. Give each child a copy of the 'Targets' photocopiable sheet on page 43 to support their calculations.
More confident: challenge the children to make as many target numbers as they can, using the three number cards, before revealing the actual target number.

Key questions

- How did you work out your answers? Did this method help you to 'hit the target' each time?
- What tips would you give somebody who was new to the game?

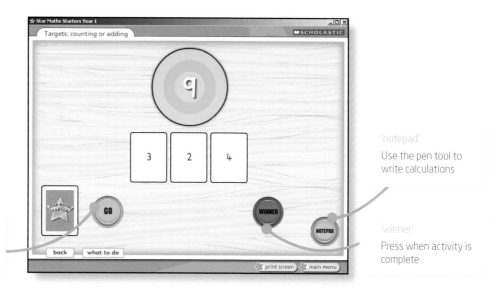

'notepad'
Use the pen tool to write calculations

'go'
Press to generate number cards

'winner'
Press when activity is complete

Making patterns

Strand

Using and applying mathematics

Learning objective

Describe simple patterns and relationships involving numbers or shapes; decide whether examples satisfy given conditions

Type of starter

Read

Whiteboard tools

- Select a length of 18, 20 or 24 beads from the 'options' menu using the on-screen keypad.
- Choose different colours from the colour palette and use to colour beads, building up a short pattern.
- To remove a colour, select white from the colour palette and press the coloured bead.
- Press 'clear' to remove all colours from all beads.

What to do

This activity encourages children to identify and make colour patterns with a simple string of beads. Create a pattern using two or three colours. Tell the children that you want to make a string of beads and you are going to use and repeat this pattern for all of the beads on the string. Ask the children to predict what will be the colour of the next bead on the string and then demonstrate this on the board. Continue for another two or three beads. Next, ask the children to continue the pattern using squared paper, or challenge individual children to come to the board to colour each bead. Repeat the activity, starting with different patterns and/or lengths of beads. On occasion, put in an 'incorrect' colour and prompt the children to check if the pattern is correct.

Differentiation

Less confident: use two colours only, so that children get used to the idea of a repeating pattern.
More confident: use three colours and one colour more than once, to extend thinking skills. Encourage the children to change the shape of the bead string, using squared paper (for example, from a line to a square).

Key questions

- *What colour is the next bead on the string?*
- *How would you describe the pattern of coloured beads?* (Using ordinal numbers such as 'first', 'second', 'third' and so on.)

colour palette
Select different colours for pattern

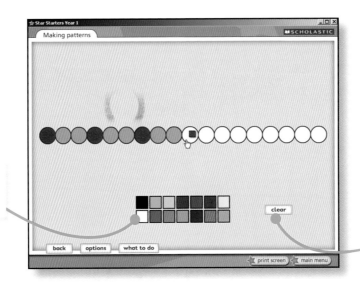

'clear'
Press to remove bead colours

Twenty cards (ITP): ordering

Strand

Counting and
understanding number

Learning objective

Compare and order
numbers, using the
related vocabulary

Type of starter

Refine

Whiteboard tools

● Select 'make stack' and set the following: 'How many cards': 5; 'First card number': 1;
   'Step number': 1; 'Step increment': 0. Press 'go'.
● Press the red 'deal' button to deal the stack of number cards in a straight line, in order.
● Press the red 'spread' button to spread the number cards around the screen.
● Make individual number cards and spread them around the screen.
● Turn a card by pressing on the red area.
● Move a card by pressing on the centre arrow and dragging it.

What to do

Use this ITP to refine the ordering of numbers through number cards up to 20, as well
as to rehearse counting in steps of 2. The ITP generates individual number cards and
stacks (decks) of number cards, which appear face down. Drag number cards around
the screen and press to turn them over and display the numbers. When making a stack,
keep the increment to 0 in order to reveal a standard deck (as in the image below
for adding in twos). Experiment with producing sets of cards in a sequence, or as a
random set.

Use the ITP to display groups of different numbers that the children can compare
and order. The stacks or spreads created can support work on identifying, describing,
extending and generating sequences. Turn over several cards in a set and challenge
the children to say what the face-down cards are. Check if the children can do this by
counting, or by recognising the pattern or sequence.

Differentiation

Less confident: to check understanding, ask the children to re-order a completed
number sequence by pressing the 'shuffle' button.
More confident: extend the number sequences above 20 and ask the children to
predict what the next number is, then what the pattern might be.

Key questions

● *Can you continue this sequence?*
● *When we shuffle the numbers, how would you go about re-ordering them?*
● *Which numbers come before and after this number?*

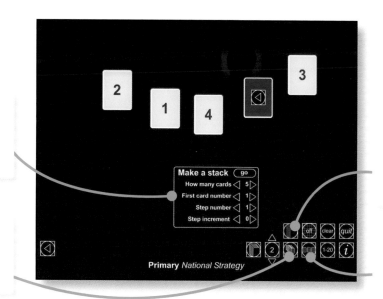

'first card number'
You can make a card
with any number up
to 999, but keep it
low for this activity

'spread'
Press to randomly
spread stack of cards
around the screen

'make stack'
Create stacks (decks)
of cards by given
criteria

'deal'
Deal a stack of cards
in a straight line,
in order

Bricks: numbers to 20

Strand

Counting and understanding number

Learning objective

Compare and order numbers, using the related vocabulary

Type of starter

Rehearse/Reason

Whiteboard tools
- Press 'go' to reveal five bricks numbered between 1 and 20.
- Drag and drop each brick into the wall, with the smallest number in the lowest position. Any bricks placed incorrectly snap back to their starting positions.
- If all of the bricks are placed correctly, a 'Well done' message appears.
- Press 'go' again to select a new set of numbered bricks.

What to do

Use this activity to rehearse ordering numbers to 20, or to test the children's understanding of numbers to 20. Press 'go' to reveal five bricks, each showing a number between 1 and 20. Ask the children to position the bricks in the wall by dragging and dropping them. The smallest number needs to be in the lowest position. Ask the children to decide on the correct order (working as a whole class, in pairs, or individually) by writing answers on their individual whiteboards. After all of the five bricks have been placed correctly, a 'Well done' message appears. Any bricks placed incorrectly will snap back to their starting positions. Discuss with the children why a brick might have been placed incorrectly, and re-order.

Differentiation

Less confident: use a number line to support the children's ordering skills, before positioning the bricks in the wall.
More confident: ask the children: *What would need to be added to the top brick to make 10 or 20?*

Key questions

- *Which are the smallest and largest numbers? Where would we place this brick?*
- *What would the new number be if 1, 2 or 3… were added to the number on the lowest brick?*

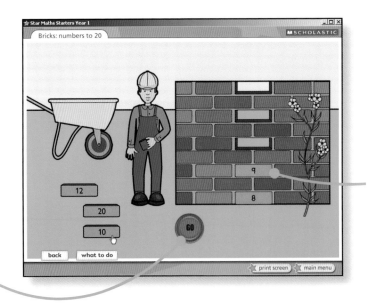

'go'
Press to generate a new set of number bricks

bricks
Drag and drop bricks into the wall

Dominoes: numbers and names

Whiteboard tools

- Press 'new' to start a new game.
- Press 'miss a go' to take another domino from the pot.
- Drag and drop the dominoes into the playing space. Rotate each domino through 90° by pressing the top right-hand corner.
- Press 'winner' if Player 1 or Player 2 has placed all of the dominoes, and it is agreed that the last domino was placed correctly.

What to do

The aim of this activity is to match the domino numbers with the appropriate written number words. The game is played in the same way as standard dominoes, with two players (or teams) playing against each other. Each player (or team) is dealt four dominoes. The remaining dominoes are left in a central pot. A starter domino is automatically generated by the computer to begin the game, and the players then take turns to play. If a player is unable to place a domino, they should press 'miss a go' and take one from the central pot. The game continues until a player has placed all of their dominoes, and is declared the winner, or there are no more dominoes left in the pot. It is possible for a stalemate situation to occur, in which neither player is able to put down a domino and the pot is empty. In this case, the player with fewer remaining dominoes is declared the winner.

Differentiation

Less confident: let the children use talk partners to discuss moves, which will help to boost confidence and affirm their decisions.
More confident: play 'beat the teacher', in which children pit themselves against an adult in the classroom.

Key questions

- *How can we identify which dominoes to select? Which numbers were easy to match? Which were difficult? Why?*
- *Which domino matches the number 7? Which domino matches 9?*

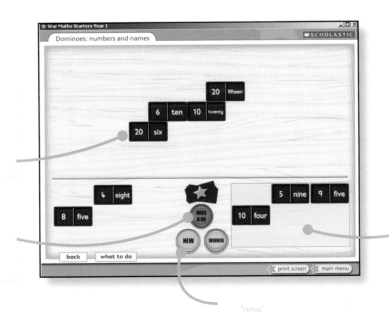

domino
- Drag domino to playing space
- Rotate by pressing top right-hand corner

'miss a go'
Press to take another domino from pot

players 1 and 2
Panel turns green to indicate whose turn it is

'new'
Press to start new game

Number line: 1 more, 1 less

Strand

Counting and understanding number

Learning objective

Say the number that is 1 more or less than any given number up to 10

Type of starter

Recall

Whiteboard tools

● Press 'hide' and then press individual numbers to hide all numbers on the line.
● Press 'clear' to reveal one hidden number.
● Press 'highlight' and then press individual numbers to highlight answers or numbers in a sequence.
● Press 'reset' to clear all highlighted or hidden numbers.

What to do

Before the lesson, hide all but one of the numbers on the number line. Ask the children, as a whole class or individually, the numbers that are 1 more and 1 less than the number shown. Use the 'clear' button to reveal the children's answers. Continue up or down the line until all the numbers have been revealed. Repeat the activity, starting with a different initial number. You can also ask the children to count in twos up to 10, hiding selected numbers along the line. Next, hide four or five numbers on the line and ask the children to identify all of them. Ask questions such as: *Where does 7 go? And 3? What is the third number I have covered on the number line?*

Differentiation

Less confident: start by revealing all the numbers and practising how to find 1 more and 1 less.
More confident: ask the children what the number would be if 10 were added to the starting number (rather than 1).

Key questions

● *What is the number that is 1 more than the one I am pointing to?*
● *What is the number that is 1 less than the one I am pointing to?*

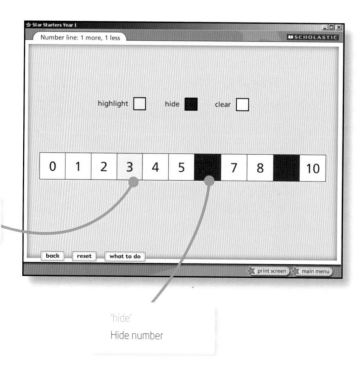

'highlight'
Highlight number

'hide'
Hide number

Counting on and back (ITP)

Strand

Counting and
understanding number

Learning objective

Say the number that is
1 more or less than any
given number, and 10
more or less for multiples
of 10

Type of starter

Refine

Whiteboard tools

- The upper '00' button shows or hides the number of beads on the left.
- The lower '00' button shows or hides the bead number.
- Press a bead to move one or several beads along the string.
- Press '-/+1' or '-/+10' to move groups of beads.

What to do

This ITP refines the basic skills of addition and subtraction, using a 100-bead string as support. The beads are coloured yellow and white, in sets of 10, to aid grouping. You can move all of the beads to the right of the string and individual (or groups of) beads to the left, by clicking on selected beads. By default, the mouse control will show the bead number as you move over it, but this function can be turned off as required. The number of beads on the left of the string is automatically shown too, but again this function can be turned off.

Practise addition and subtraction, in single numbers or multiples of 10, by moving 1 or 10 beads to the left, or right, of the string. You can also use this ITP to model different counting and calculation strategies (for example, counting on, or back, to the nearest 10).

Differentiation

Less confident: encourage the children to count each bead as it is moved along the string. If available, use 100-bead or 10-bead strings in class. The children could also make their own bead counters for counting at home or school.
More confident: increase the increment steps. For example, can the children predict what is 20 more than 24, using the beads to support their thinking?

Key questions

- *Why are the beads in different colours? How can this help us to count?*
- *What is 1 more than ___? What is 10 more than ___? Can you show me your answer using the beads?*

beads
Press to move one
or several beads
along string

large circles at either
end of string
Press to move
all the beads to
opposite end

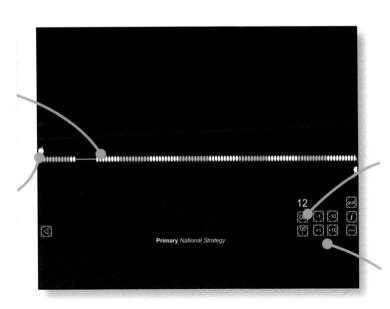

Primary *National Strategy*

'00' button
Press to show or hide
the number of beads
moved to the left of
the string

'-1/10' and '+1/10'
buttons
Press to change the
number of beads on
the left in steps of
1 or 10

Bricks: multiples of 10

Strand

Counting and understanding number

Learning objective

Say the number that is 10 more or less for multiples of 10

Type of starter

Rehearse/Reason

Whiteboard tools

- Press 'go' to reveal five bricks numbered between 1 and 100.
- Drag and drop each brick into the wall, with the smallest number in the lowest position.
- After all bricks are placed correctly, a 'Well done' message appears.
- If any bricks are placed incorrectly, they snap back to their starting positions.
- Press 'go' again to select a new set of number bricks.

What to do

Use this activity to rehearse their knowledge of multiples, or to test their understanding of multiples of 10. Press 'go' to reveal five bricks, each showing a multiple of 10 between 10 and 100. Ask the children to position the bricks in the wall by dragging and dropping them into the appropriate spaces.

The smallest number needs to be in the lowest position. Ask the children to decide on the correct order (working as a whole class, in pairs, or individually) by writing answers on their individual whiteboards. If all bricks have been placed correctly, a 'Well done' message appears. If any have been placed incorrectly, they snap back to their starting positions. Discuss with the children why a brick might might have been placed incorrectly, and re-order.

Differentiation

Less confident: use a number line to support the children's ordering skills, before positioning the bricks in the wall.
More confident: ask the children: *What would need to be added to the top brick to make 100, or even larger numbers?*

Key questions

- *Which are the smallest and largest multiples of 10 in this wall?*
- *What would the new number be if 1, 2 or 10… were added to the number on the lowest brick?*

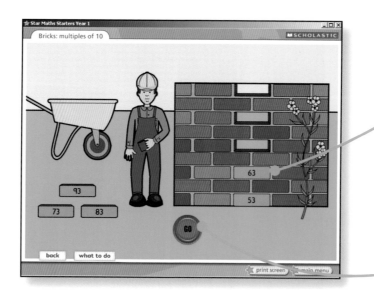

bricks
Drag and drop bricks into the wall

'go'
Press to generate a new set of number bricks

Bingo: number bonds to 10

Strand

Knowing and using number facts

Learning objective

Derive and recall all pairs of numbers with a total of 10

Type of starter

Rehearse

Whiteboard tools
- Press 'set timer' to adjust the time between bingo calls (5 to 20 seconds).
- Press 'start' to start a new game.
- Press 'check grid' to check answers.
- Press 'play on' or 'winner' after checking a player's grid.
- Press 'new game' to start a new game.

What to do

The aim of this activity is for children to rehearse their knowledge of pairs of numbers that total 10 against a time limit. Ask the children to play individually, or organise them into pairs. Provide each child (or pair) with a bingo card. The bingo card template can be printed from the opening screen of the CD-ROM, or photocopied from page 44.

Each bingo ball offers a single-digit number (including zero). The children need to find a number on their bingo cards which, when added to the number on the bingo ball, makes a total of 10. Once found, ask the children to mark this number on their cards. If the children are new to the game, allow for a longer amount of time between bingo calls. If a child calls *House* (or any other similar winning call you have chosen), press 'check grid' to pause the game and view the checking grid, which includes all of the completed number sentences that have been created. If the child is correct, press the 'winner' button for a fanfare, or press 'play on' to continue the game.

Differentiation

Less confident: support the children with counting blocks. Work with the children on establishing suitable mental methods to improve recall.
More confident: increase the number of answers on the bingo cards, using the bingo card template on page 44. To challenge the children, reduce the time allowed between calls on the timer.

Key questions
- *How can we check that we have all the correct answers?*
- *What strategies can we use to remember these number facts?*

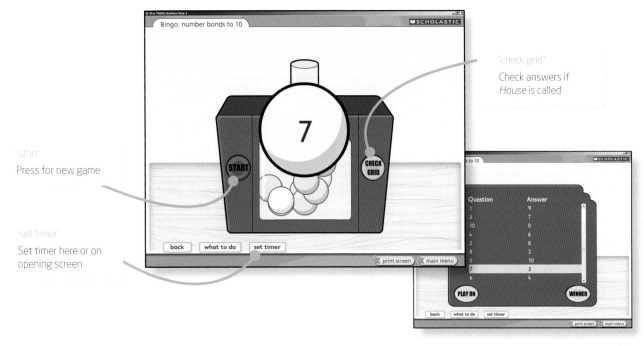

'check grid'
Check answers if *House* is called

'start'
Press for new game

'set timer'
Set timer here or on opening screen

Addition square: addition facts to 5

Strand

Knowing and using number facts

Learning objective

Recall addition facts for totals to at least 5

Type of starter

Recall

Whiteboard tools

- Press 'options' and set the following: 'squares on a side': 10; 'start number': 1; 'step': 1.
- Press 'hide' to hide all the totals of 11 or more.
- Use the 'highlight' button to demonstrate calculations.
- Use the 'clear' button to clear selected highlighted or hidden numbers.
- Press 'reset' to clear all highlighted or hidden numbers from the grid.

What to do

The aim of this starter is for children to recall addition facts to 5 and above. Prepare the addition square before the lesson. Hide all the totals in the table with 11 or more. Choose a number below 10 (for example, 5). Ask the children to tell you all the calculations that they know to make 5 using addition, such as 1 + 4, 4 + 1, 3 + 2, and so on. Concentrate on pairs of numbers (rather than, for instance, 1 + 1 + 3). Once a calculation has been given, use the 'highlight' button to mark the calculation on the square. Continue until all those possible combinations for the number 5 are exhausted. In the same way, challenge the children to offer addition facts for other totals (up to 5 initially, and then to 10) until all totals in the square, apart from those that were originally hidden, have been highlighted. You will need to clear some or all of the hidden boxes to extend the activity in this way.

Differentiation

Less confident: focus on addition facts to 5 and support children with their calculations if necessary.

More confident: extend the number square to totals of 15 or more. Ask the children to mark off calculations as they identify them.

Key questions

- *What are the possible addition calculations that make 5? Is 1 + 4 the same as 4 + 1?*
- *What other ways are there to make 5, 6, 7...?*

'hide'
Press to hide squares or columns

'options'
Specify size and steps of square

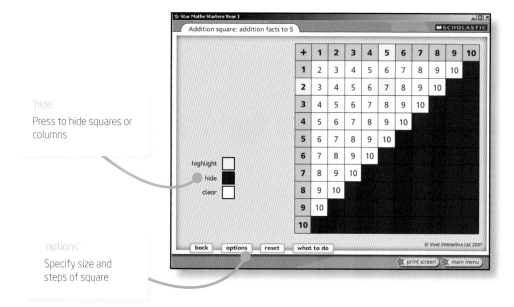

Number spinners (ITP)

Strand

Knowing and using number facts

Learning objective

Derive and recall all addition facts for totals to at least 5; work out the corresponding subtraction facts

Type of starter

Refine

Whiteboard tools

- Press the up and down arrows on the 'range' button to select the range of numbers shown on the spinners.
- Press the 'shape' button arrows to select three-, four-, five- or six-sided number spinners.
- Press the up and down arrows on the 'spinner' button to select the number of spinners displayed (from one to three).
- Spinners: press the central yellow dot to spin the spinners; press a number on the spinner to increase it by 1, within the number range selected.

What to do

The aim of this ITP is to generate random numbers, using 'spinners' with three, four, five or six sides. Using the up and down arrows on the 'spinner' button, you can create between one and three spinners at a time. The arrows on the 'shape' button allow you to select the number of sides the spinners will have. Once this has been determined, press the solid white shape inside the 'shape' button to display the spinners. Press the central yellow dot on the spinner to spin it to generate a random number (as indicated by the red arrow beneath the spinner).

Use this ITP to create incomplete number sentences to rehearse number bonds up to 5 (though it can quickly and easily be extended to 10). For example: 2 + 2 + ? = 5. Alternatively, one spinner could be used to generate a single figure and the class has to provide the other.

Differentiation

Less confident: use two spinners to add three random numbers together. In this case, reduce the number range to up to 2 and use a three-sided spinner.
More confident: select a higher number range (up to 10 or 20, for example) and spin two numbers where the range is up to 5 or 10.

Key questions

- *What is the missing number?*
- *How could this be written as an addition or subtraction sentence?*

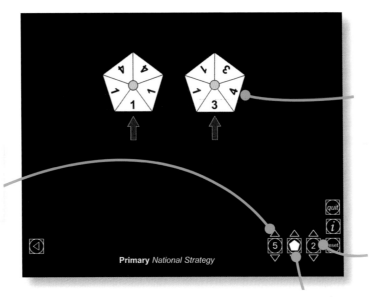

'spinners'
- Press the yellow dot to spin the spinner
- Press any individual number to increase it by 1

'range' button
Press arrows to select the range of numbers (0–20)

'spinner' button
Press to create the number of spinners (1-3) required

'shape' button
Press to choose spinner shape

Primary National Strategy

Beanstalk: addition and subtraction

Strand

Knowing and using number facts

Learning objective

Derive addition facts for totals to at least 5; work out the corresponding subtraction facts

Type of starter

Recall

Whiteboard tools

● Press 'options' to select 'up only' (addition) or 'up and down' (addition and subtraction), and whether to show or hide the number sentence.
● Drag Jack up and down the beanstalk, in steps of 1, to answer the number problem.
● Use the 'pen' tool in the 'notepad' to show each calculation, before moving Jack up and down the beanstalk, or to write your own number problem.
● Press 'new' to move Jack back to 0 and to generate a new number problem.
● Press 'answer' to show the completed number sentence.

What to do

Use this activity to practise the addition of numbers to 11 and subtraction of numbers to 12. The activity sets a range of number problems, to be solved using a 'Jack and the Beanstalk' theme. To support the children with their calculations, print the beanstalk template from the CD-ROM, or photocopy it from page 45. Encourage the children to work out the answers for themselves, using their individual whiteboards, before any class discussion. Extend the activity beyond the initial question by asking, for example: *What would happen if Jack now moved down another 2?* Illustrate this by dragging Jack to the new position.

Differentiation

Less confident: use the 'up only' option at first, to focus on addition facts. Alternatively, provide your own questions to limit the number range in each session. Provide a copy of photocopiable page 45 to support the children's calculations.
More confident: ask additional questions that would move Jack beyond the number 12.

Key questions

● *On which number does Jack end up? How do we know?*
● *How far up or down the beanstalk would Jack need to climb to reach* (for example) *the number 5?*

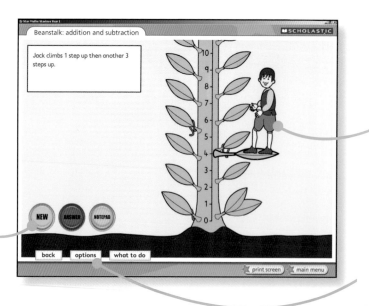

'new'
Press to generate a new problem

Jack
Drag Jack up or down the beanstalk

'options'
● Select 'up only' or 'up and down'
● Show or hide the number sentence

Sorting machine: count in twos and fives

Knowing and using number facts

Learning objective

Count on in twos and fives and use this knowledge to derive the multiples of 2 and 5

Type of starter

Reason

Whiteboard tools
- Press 'go' to launch the first number ball into the sorting machine.
- Press one of the 'cogs' to decide whether the number is a multiple of 2 or a multiple of 5.
- Press 'reset' to bring up a new set of four number balls.

What to do

The aim of this activity is for the children to work out how to sort numbers into multiples of 2 and multiples of 5. Press 'go' to launch the first number ball into the sorting machine and ask the children to vote whether they think the number is a multiple of 2 or a multiple of 5 (either by a show of hands, or by writing on their individual whiteboards). Check by asking the children to count up in twos or fives. In subsequent sessions, ask individual children to give reasons for their selections, using appropriate mathematical language. Focus, in particular, on numbers that are multiples of 2 and 5 (tens numbers).

Differentiation

Less confident: provide the children with a number line for support. Ask another adult to support the children in counting up to the target number.
More confident: encourage the children to predict which 'bucket' the number should go into when it first appears. Can they identify any numbers that are multiples of 2 and 5?

Key questions

- *Why did you sort the number in this way? How can you spot that this is a multiple of 2 or 5?*
- *What other multiples of 2 or 5 can you think of?*

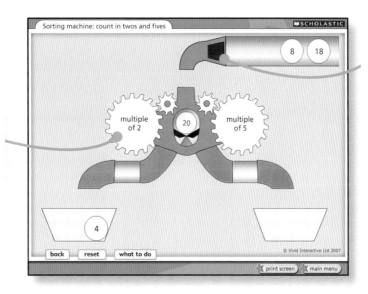

cogs
Press to sort multiples

'go'
Press to launch number ball

Number square: ones, twos, fives and tens

Strand

Knowing and using number facts

Learning objective

Count on or back in ones, twos, fives and tens and use this knowledge to derive the multiples of 2, 5 and 10

Type of starter

Recall

Whiteboard tools

- Press 'options' and select the following: 'squares on the side': 10; 'start number': 1; 'step': 1.
- Use the 'hide' button to hide all of the columns except the first, second, fifth and tenth.
- Select 'clear' and press on any hidden square to reveal the number beneath.
- Use the 'highlight' button to highlight any number on the grid.
- Press 'reset' to clear all highlighted or hidden numbers on the grid.

What to do

Prepare the number square before the lesson (see image below). Ask the children to count on and back in ones, while pointing to the first column. Next, ask them to count on and back in twos, while pointing to the second column. Move on to derive multiples of 2. Highlight one of the squares in the second column (for example, 8) and ask the children questions such as: *How many twos make 8? What number multiplied by 2 gives 8?* Once the children are confident in using the number 2, continue this activity using the columns for numbers 5 and 10. Initially, it might be better to use this starter for each of the numbers 2, 5 and 10 separately before combining multiples in subsequent sessions.

Differentiation

Less confident: invite the children to count on and back; initially, focus on multiples of 2.
More confident: ask the children to tell you a number sentence associated with each multiple.

Key questions

- *How many twos make 4? What is 10 more than 4?*
- *Who can give me a multiple of 5? A multiple of 10? How do we know they are multiples of 5 or 10?*

'highlight'
Press to highlight any square

'hide'
Press to hide squares or columns

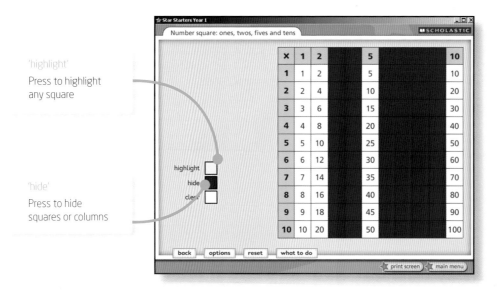

Addition square: doubles to 10

Strand

Knowing and using number facts

Learning objective

Recall the doubles of all numbers to at least 10

Type of starter

Recall

Whiteboard tools

● Press 'options' and set the following: 'squares on a side': 10; 'start number': 1; 'step': 1.
● Use the 'highlight' button to pick out all the diagonal numbers (the doubles).
● Use the 'hide' button to hide all the doubles.
● Select 'clear' and then press a hidden number to reveal it.
● Press 'reset' to clear all highlighted or hidden numbers.

What to do

Prepare the addition square before the lesson. Highlight the doubles and ask the children to say these aloud up to double 10 (20). Next, hide all of the doubles and challenge the children to answer some quick-fire questions, such as: *Double 4? Half 10? Twice 3?* and so on. After each selection, clear the square to confirm each answer. Ask the children to write the answers on their individual whiteboards or invite individual children to come to the board to reveal the answer on the addition square.

Differentiation

Less confident: give the children individual number lines, or ask them to focus on numbers surrounding the doubles on the number square, to help them.
More confident: move the children on to doubles and halves above 10.

Key questions

● *What is the double of this number?*
● *What is half of this number?*

'hide'
Press to hide squares or columns

'clear'
Press to reveal number of square

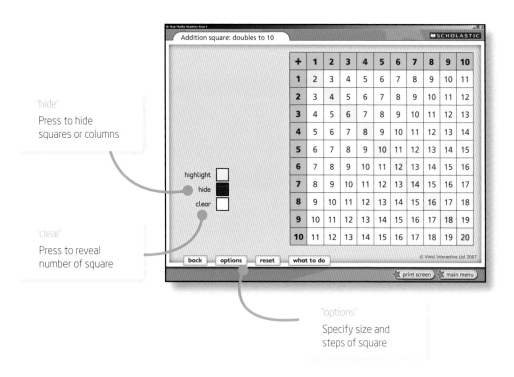

'options'
Specify size and steps of square

Bingo: doubles

Strand

Knowing and using number facts

Learning objective

Recall the doubles of all numbers to at least 10

Type of starter

Rehearse

Whiteboard tools

- Press 'set timer' to adjust the time between bingo calls (5 to 20 seconds).
- Press 'start' to start a new game.
- Press 'check grid' to check answers.
- Press 'play on' or 'winner' after checking a player's grid.
- Press 'new game' to start a new game.

What to do

The aim of this activity is for children to rehearse their knowledge of doubles, up to double 15, against a time limit. Ask the children to play individually, or organise them into pairs. Provide each child (or pair) with a bingo card, which can be printed from the CD-ROM, or photocopied from page 44.

Each bingo ball offers a question on doubles or halves. The children need to work out the answer and mark the number if it appears on their bingo cards. If the children are new to the game, allow for a longer period of time between bingo calls. If a child calls *House* (or any other winning call you have chosen), press 'check grid' to pause the game and view the checking grid, which includes all of the completed questions. If the child is correct, press the 'winner' button for a fanfare, or press 'play on' to continue the game.

Differentiation

Less confident: support the children with counting blocks. Work with them to refine mental strategies for working out doubles.
More confident: increase the number of answers on the bingo card, using the bingo card template on page 44.

Key questions

- *How can we check that we have all the correct answers?*
- *What strategies can we use to learn doubles more quickly?*

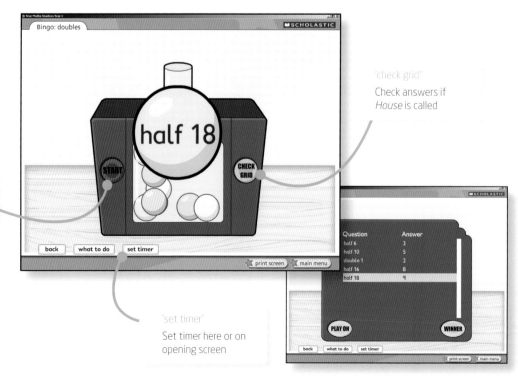

'start'
Press for new game

'check grid'
Check answers if *House* is called

'set timer'
Set timer here or on opening screen

Maths Boggle: adding numbers 1 to 6

Strand

Calculating

Learning objective

Recognise that addition can be done in any order

Type of starter

Rehearse

Whiteboard tools

- Use 'options' to select a target question.
- Press the dice to highlight them.
- Use the 'notepad' to show number sentences.
- Press 'new' to rattle the Boggle dice.

What to do

This activity is designed to show the children that addition can be done in any order, using the Boggle dice to answer set questions.

Start by selecting a question from the 'options' menu at the bottom of the screen (or, should you wish, by setting your own question). Rattle the dice by pressing 'new' to reveal a random selection of single-digit numbers. Ask the children to work in pairs or individually, to find the answer to the question using only the numbers on the screen. Encourage them to discuss their answers with their partners and to write the answers on their individual whiteboards. Highlight the dice to help the children to visualise the answer. Use the pen tool from the on-screen notepad to show calculations, or to write your own number sentence.

Differentiation

Less confident: using 'options', set the target question 'add two numbers together' and focus initially on using the lowest numbered dice.

More confident: after the children have found the initial answer, ask supplementary extension questions, such as: *Now add 5* or *Now add the largest number you can see.*

Key questions

- *Is 1 + 4 the same as 4 + 1?*
- *Add the largest number you can see to the smallest number you can see. Add the smallest number you can see to the largest number you can see.*

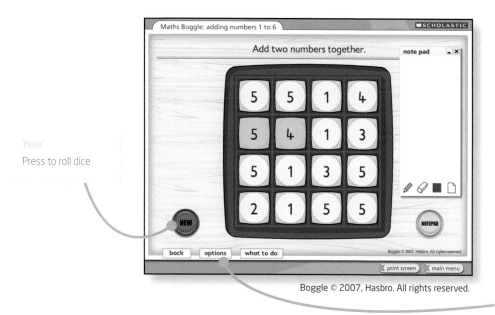

'new'
Press to roll dice

'options'
Press to set target question

Boggle © 2007, Hasbro. All rights reserved.

Number facts (ITP)

Strand

Calculating

Learning objective

Understand subtraction as 'take away' and find a 'difference' by counting up

Type of starter

Refine

Whiteboard tools

- Use the up and down arrows on the 'number' button at the foot of the screen to select the number of beads (maximum of 20). Press the number to display the beads.
- Select '+' or '-' to switch between addition and subtraction sentences.
- Press the '? + ? =' button to show or hide the number sentence on the screen.
- Press '-' and select one of the options above, to show either an opaque or translucent container.
- Drag and drop beads into the container to demonstrate the pattern of 'taking away'.

What to do

This ITP allows the children to refine their understanding of subtraction, by 'taking away' beads and placing them in a container to demonstrate different subtraction sentences. The activity can also be used to recall addition and subtraction facts.

The ITP displays up to 20 coloured beads and the corresponding addition or subtraction number sentence. Select 10 beads and the subtraction option ('-'); 10 pink beads and a container then appear on the screen. The sentence '10 - 0 = 10' also appears on the screen. Drag one bead into the container and the sentence changes to '10 - 1 = 9'. Remind the children that you have 'taken away' one bead to leave 9. Continue by pressing '? + ? =' to hide the number sentence. Take away four beads and ask the children to write the new number sentence on their individual whiteboards. Check their answers and then reveal the number sentence on the board. With the next examples, make the box opaque to hide the subtracted beads. Press the subtracted figure in the number sentence to change it to '?'. Then reveal the beads to check the children's answers.

Differentiation

Less confident: subtract one bead at a time and do not hide the number sentence, to support the children in understanding 'taking away'.
More confident: extend the activity to '20 - 0' and hide each number sentence.
Invite volunteers to come to the board and make particular sentences by adding or subtracting beads from the container.

Key questions

- *How many beads would we need to take away to make this subtraction sentence?*
- *How would you write the following subtraction sentence?*

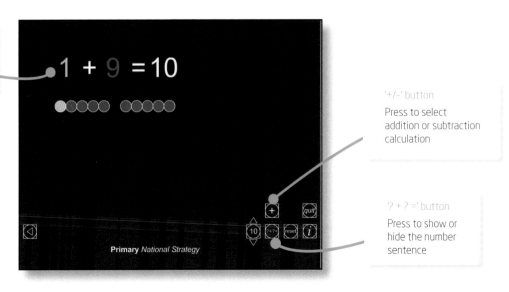

number sentence
Press each number to replace it with a question mark

'+/-' button
Press to select addition or subtraction calculation

'? + ? =' button
Press to show or hide the number sentence

Number line (ITP)

Strand

Calculating

Learning objective

Find a 'difference' by counting up

Type of starter

Refine

Whiteboard tools

- Use the arrows on the 'min' and 'max' buttons to increase or decrease the range of numbers on the number line.
- Drag the marker and drop to place second marker.
- Press the blue handlebar button to link the markers and show the difference span.
- Press the '000' buttons to show/hide the numbers in the spanning boxes.
- Press '? + ? =' to show/hide the number sentence.

What to do

The aim of this starter is to model and refine calculation strategies for subtraction, and to develop the children's understanding of finding differences, using a number line. This is a versatile ITP and can quickly be adapted to increase the range appropriately.

To start with, move the two markers along the number line to generate a number sentence. Depending on the order of the two markers, you can show the sum and difference between the two numbers and the calculation represented. Go through a few examples with the number in the spanning box shown and then challenge individual children to work out number sentences with the number in the spanning boxes, or the complete number sentence, hidden. A subtraction sentence can be made by dragging the yellow marker across the blue marker.

Differentiation

Less confident: limit the range to 0–10 and ask the children to count along the number line to find the differences.

More confident: extend the number range to 0–20, using the pointers alongside the 'max' button, or show just the answer on the number line (hide the span). Ask: *What other pairs of numbers could be used to equal that answer?*

Key questions

- *What is the difference between 4 and 6? How can you show this on the number line?*
- *What other 'difference numbers' will give us the same answer?*

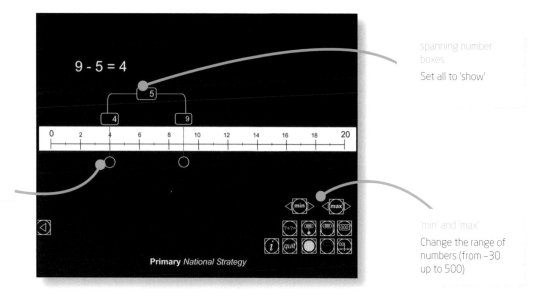

circle
Drag circle to change numbers at any point on the line

spanning number boxes
Set all to 'show'

'min' and 'max'
Change the range of numbers (from –30 up to 500)

Function machine: addition and subtraction

Strand

Calculating

Learning objective

Use the vocabulary related to addition and subtraction to describe number sentences

Type of starter

Read/Reason

Whiteboard tools

- Press 'options' and select 'random' mode for a set of computer-generated number sentences. The 'input', 'output' and 'function' windows will be closed initially. Any window can be opened by pressing it.
- Press 'options' and select 'manual' mode to input your own number sentences.
- Press 'history' to view a list of all number sentences completed during the lesson.
- Press 'go' to open all windows and reveal the completed number sentence.
- Press 'new' to start a new sentence.

What to do

The aim of this activity is to find missing numbers, or the function, to complete a number sentence. You can input your own number sentences ('manual' mode) or allow the computer to generate them ('random' mode).

Random mode: the computer generates a number sentence, but hides the input, output and function windows on the machine. Decide which window to open first and, after one other element has been revealed, ask the children to write the complete number sentence on their individual whiteboards. Check their answers and then press 'go' to reveal the number sentence. The computer will generate both addition and subtraction sentences with answers from 0 to 10.

Manual mode: enter some number sentences involving the addition or subtraction of pairs of numbers (for example: 1+ 9, 3 + 2, 7 - 3). Press 'history' to check answers.

Differentiation

Less confident: hide the input number and function to show just the answer. Ask the children to think of a number sentence which will give that answer. Narrow the options down further, by revealing the input number or function.

More confident: use the machine to demonstrate sequences. For example, model adding 2 to a number by keeping the function to +2 and keying the created output number back into the input box, (a record of this will be kept in the 'history'). Extend the number range in 'manual' mode.

Key questions

- *How did you work out the missing part of the sentence?*
- *How much of the sentence needs to be revealed before you can complete it?*

'new'
- Press to start again in 'manual' mode
- Press for number sentence in 'random' mode

'options'
- Select 'manual' mode' to input numbers manually
- Select 'random' mode" for computer-generated numbers, initially hidden

'history'
Press to view list of completed sentences

'go'
Press to reveal answer

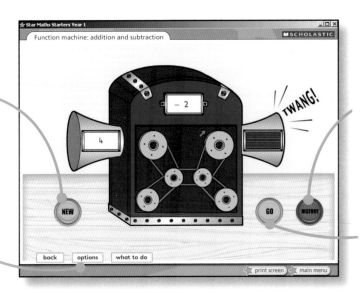

Number sentences: missing numbers

Strand

Calculating

Learning objective

Use the vocabulary related to addition and subtraction and symbols to describe and record addition and subtraction number sentences

Type of starter

Read

Whiteboard tools

- Drag and drop number and symbol cards onto a line to build a number sentence.
- Drag and drop numbers and symbols within a line to re-order them.
- Drag numbers and symbols off the line to remove them.
- Press 'reset' to clear the lines and start again.

What to do

This activity encourages children to read number sentences as well as do the calculations. Create a number sentence and then ask the children to read it out. Numbers of any size can be selected, as two digits selected consecutively snap together to form a two-digit number. Ask the children for the answer to the calculation and then ask them to read the number sentence again, inserting the previously missing number. Initially, use numbers up to 10 and prepare number sentences that are sequential, such as 1 + 1= □, 1 + 2 = □, 1 + 3 = □, or 8 - 1= □, 8 - 2= □, so that reading the complete sentence is emphasised and practised, rather than just focusing on the answers.

Differentiation

Less confident: check that the children understand the meaning of the symbols before starting the activity. Limit the number range to addition and subtraction facts to 5.
More confident: ask the children to come to the whiteboard to make their own number sentences for the whole class to answer.

Key questions

- *What is this number sentence? What is this (+) symbol?*
- *How would you show 4 + 2 = 6 on the whiteboard?*

number sentence
Drag numbers and symbols to re-order them

'reset'
Press to clear screen

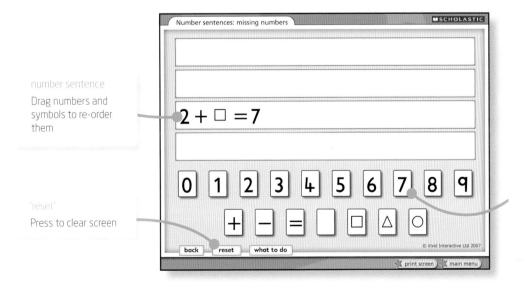

cards
Drag numbers and symbols to a line to build a number sentence

Shopping: money problems

Strand

Calculating

Learning objective

Solve practical problems that involve combining groups of 2, 5 or 10

Type of starter

Refine

Whiteboard tools

● Drag an item into the shopping basket.
● Press 'check-out' to move to next screen.
● Drag money into the till to pay for the item. Press 'sale' to check if the amount paid is correct.
● Press 'clear' to empty the till.
● Press 'back to shop' to select a new item.

What to do

This activity is designed to introduce children to coins of different values up to 20p. Invite a volunteer to come to the whiteboard to select an item from the shop and then drag and drop it into the shopping basket. Next, challenge the class to work out which coins you would need to use to pay for the item. Drag and drop coins from the purse into the till (all coins can be used more than once). Press 'sale' to check that the amount paid is correct. Focus on the different combinations of coins required to pay for each item and prompt for alternatives (for example, to pay for the crisps – priced 10p – the children could use combinations including 1 × 10p coin, 2 × 5p coins or 1 × 5p coin, 2 × 2p coins and 1 × 1p coin).

Differentiation

Less confident: select the items with lower prices and those that require fewer coins, to build confidence. Allow the children longer to work out the coins needed. Provide some play money (or real money) to reinforce the children's understanding.
More confident: challenge the children to find all the different combinations of coins they could use to pay for an item.

Key questions

● *Which coins did you use to pay for the item?*
● *Could we pay using a different set of coins?*

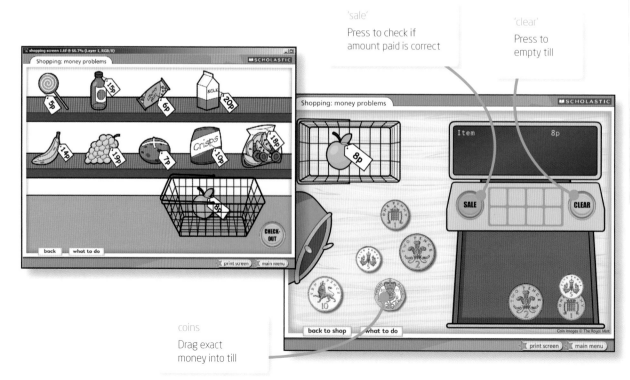

'sale'
Press to check if amount paid is correct

'clear'
Press to empty till

coins
Drag exact money into till

Dominoes: 2D shapes

Strand

Understanding shape

Learning objective

Visualise and name
common 2D shapes

Type of starter

Refresh/Read

Whiteboard tools
● Press 'new' to start a new game.
● Press 'miss a go' to take another domino from the pot.
● Drag and drop the dominoes into the playing space. Rotate each domino through 90° by pressing the top right-hand corner.
● Press 'winner' if Player 1 or Player 2 has placed all of the dominoes, and it is agreed that the last domino was placed correctly.

What to do

The aim of this activity is to refresh children's understanding of common 2D shapes and to help them visualise the shapes by matching the shapes to their written names. The game is played in the same way as standard dominoes with two players, or teams, playing against each other. Each player (or team) is dealt four dominoes, with the others left in a central pot. A starter domino is selected by the computer to begin the game, and the players then take turns to play. If you are placing the dominoes for the children, make sure at all times that the children use correct mathematical vocabulary to describe the shapes they wish to place. If a player is unable to place a domino, they should presss 'miss a turn' and take one from the central pot. The game continues until a player places all of their dominoes, and is declared the winner, or there are no more dominoes left in the pot. It is possible for a stalemate situation to occur, in which neither player is able to put down a domino and the pot is empty. In this case, the player with fewest remaining dominoes is declared the winner. Encourage the children to describe the shapes on the domino before placing each domino.

Differentiation

Less confident: let the children use talk partners to discuss moves, which will help to boost confidence and affirm their decisions. Provide some flat shapes to further support their understanding.
More confident: play 'beat the teacher', in which children pit themselves against an adult in the classroom.

Key questions

● Which shapes were easy to spot? Which were difficult? Which shape matches the word 'triangle'?
● Can you find a domino shape with four corners and two long and two short sides? Where would you place it? What is its name?

domino
• Drag domino to playing space
• Rotate by pressing top right-hand corner

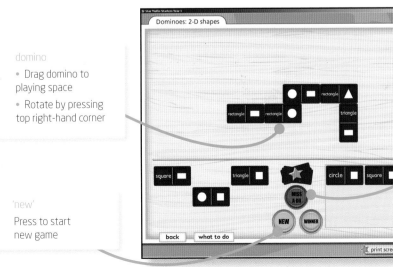

'miss a go'
Press to take another domino from the pot

'new'
Press to start new game

players 1 and 2
Panel turns green to indicate whose turn it is

Maps and directions

Strand

Understanding shape

Learning objective

Recognise and make whole, half and quarter turns

Type of starter

Reason

Whiteboard tools

- Drag and drop the direction and movement cards to prepare the route. Press 'move' to confirm the chosen route.
- Press 'show route' to display the directions selected so far.
- View the box at the top of the screen to identify which items the knight has collected along the route.
- Press 'new' to generate a new map.

What to do

This activity develops children's understanding of position and movement. From a given starting point, ask the children for directions to guide the knight through the castle, using the direction and movement cards on the screen as prompts. The knight has to collect his sword, shield and helmet before making his way to the exit. Drag and drop each card into place to build up the route. The order in which the cards are placed affects whether the knight turns first or moves first. Press 'move' after each selection.

More than one route is available and some routes include barriers – so the quickest route is not necessarily the best. The position of the items and barriers changes randomly with each game. Press the 'show route' button to check the knight's progress. Encourage the children to check and challenge the route at this stage and start again if necessary. Try to ensure at all times that they use the correct vocabulary when selecting an instruction from the screen and use other vocabulary where appropriate (for example: *Turn right towards the tent*).

Differentiation

Less confident: provide copies of photocopiable page 46 as additional support.
More confident: ask the children for a description of the route they will take before starting the activity. Let them use photocopiable page 46 to describe this route.

Key questions

- *What is the quickest way to move the knight to his shield? Out of the castle?*
- *Are there any other ways of moving from this point to this point?*

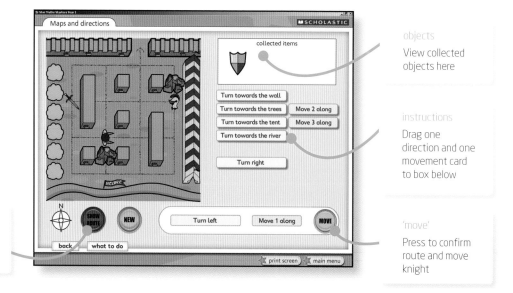

Star Maths Starters ★ Year 1

Find that cat!

Strand

Understanding shape

Learning objective

Visualise and use everyday language to describe the position of objects

Type of starter

Reason

Whiteboard tools

- Type in the 'number across', then the 'number up and down', to select a square.
- Press 'check' to confirm the choice.
- Continue to select different squares until the cat has been found.
- Press 'new' to start a new game with the cat in a different position.

What to do

The aim of this activity is to find a square on a 3 × 3 grid in which a cat is hiding. The position of the cat is randomly generated each time.

Explain to the children how they should select 'number up and down' and 'number across' to identify each square. Discuss how they might use these numbers to describe the position of different squares. After each selection, the square will be revealed. If the cat has not been found, ask the children about the remaining possible squares that the cat might be hiding. At all stages, ask the children to describe the square they are selecting as well as giving its number (for example: *the square above the watering can; the top middle square,* and so on). When the cat has been found, the garden is completed and a miaow sound is heard! Children should respond positively to this activity and will quickly learn not to repeat their selected squares.

Differentiation

Less confident: check carefully that the children understand how to identify each square.
More confident: later in the game, ask the children to identify which squares have not been chosen.

Key questions

- *How can we describe* (for example) *the square in the top-left corner?*
- *Which square is above the mole? To the right of the flower?*
- *Which squares have not yet been picked? Which squares have been picked?*

cat
Revealed when correct square is selected

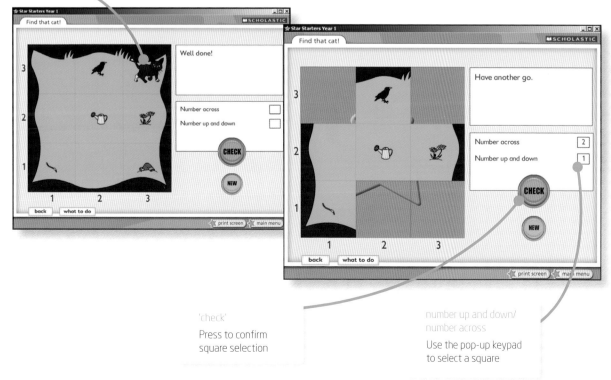

'check'
Press to confirm square selection

number up and down/ number across
Use the pop-up keypad to select a square

Pan balance: comparing weights

Strand

Measuring

Learning objective

Estimate, weigh and compare objects

Type of starter

Rehearse

Whiteboard tools

- Drag and drop items and blocks onto the pan balance.
- Remove items from the pan balance by dragging them off.
- Press 'reset' to remove all items from the pan balance.

What to do

The aim of this activity is to estimate, then compare, the weights of different objects using an on-screen pan balance. If available, this might be supported by a set of similar objects and an actual pan balance.

Explain to the children that the key question will be: *Does your object weigh more or less than mine?* Ask a child to choose an item (for example, the cheese), then ask the rest of the class whether it will weigh more or less than yours (for example, the potatoes). Is there a big difference between the two weights? Can the children estimate which of the items would weigh nearly the same? Test their predictions on the scales. Continue using the wooden blocks. Ask: *How many wooden blocks will weigh the same as the pint of milk?*

Differentiation

Less confident: give the children a range of standard classroom resources (from books to pencils and glue sticks). Can they order them from heaviest to lightest? Are there any clues to suggest what is heavy or light? (Its size, for example.)
More confident: challenge the children to start grouping some of the items that have similar weights.

Key questions

- *Is there an item that could be used to estimate the weight of other objects?* (For example: *Does a jar of jam weigh more or less than a bag or sugar?*)
- *If we know how much a small item weighs (such as a pen), can we estimate how many would weigh the same as a bag of flour?*

pan balance
Drag items or blocks on or off pan balance

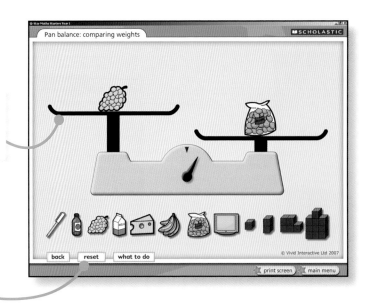

'reset'
Press to remove all items from the pan balance

Measuring jug

Whiteboard tools

- Press 'options' to set the following: 'scale mode': none; 'subdivisions': none; 'fill steps': manual.
- Press 'in' to fill the jug, and press it again to stop filling.
- Press 'out' to empty the jug, and press it again to stop emptying.
- Press 'reset' to start again.

What to do

This activity introduces measuring instruments in the form of a measuring jug. It may be helpful to label the divisions 0 to 10 with the whiteboard pen, starting with 0 at the base of the jug. Ask the children to read aloud together the divisions up to 10, as you point to them.

Tell the children that they are going to find out how much water has been put in the jug, by reading the markers on the jug. Fill the jug part way (up to, for example, the 3 mark) and then question the children about the reading. Explain that you want them to read the level to the nearest marker. Ask questions such as: *How much water would we need to put in to half fill it?* Experiment with filling and emptying the jug to find the difference between two markers, describing the levels using non-standard units such as *half full; three cups full*. Test your predictions using an actual measuring jug and other containers, such as cups, egg cups and so on .

Differentiation

Less confident: concentrate on reinforcing the children's understanding of capacity. Ask: *How many cups would fill the jug?*
More confident: challenge the children to make a more accurate reading, such as 3½ on the scale if the level falls between the third and fourth markers. Compare to reading a number line.

Key questions

- *A glass of water is poured into the jug.* (Fill jug to half way.) *How much water was in the glass?*
- *If another glass of water is added, what level will the marker be at?*

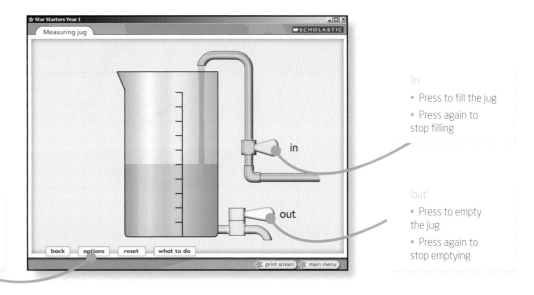

'options'
Hide the scale; select 'no subdivisions' and set 'fill steps' to manual

'in'
- Press to fill the jug
- Press again to stop filling

'out'
- Press to empty the jug
- Press again to stop emptying

Clock

Strand

Measuring

Learning objective

Use vocabulary related to time; read the time to the hour and half hour

Type of starter

Read

Whiteboard tools

● Press 'randomise' to select a random analogue time.
● Drag the clock hands to change the time on the analogue clock.
● Use the 'options' button to specify the settings for the 'randomise' functions. You can also choose to link the hour and minute hands on the analogue clock.

What to do

This starter helps the children to accurately read the time on an analogue clock face to the nearest hour and half hour. Move the clock hands by dragging them round, in the same way as with a standard classroom clock. Display a time set to the hour, such as 3 o'clock, and ask the children to role play asking for and reading the time to each other. Select another time on the hour. In subsequent sessions, ask one of the children to set the time for the others to read.

Once the children are confident with this, introduce half-hour time. (Note: if you have selected to 'link hands', when the minute hand is moved to 6 the hour hand will automatically move on a half hour-increment.) Repeat with different times. Use the randomise button to vary between times to the hour and half hour.

Differentiation

Less confident: use real-life examples to support and reinforce understanding of hourly time. Ask: *What time do we start school? What time do we have lunch?*
More confident: explain that time doesn't happen in half-hourly chunks. Ask for any examples from the day where there is a different time setting (for example, school might start at quarter to 9) and show these on the clock face.

Key questions:

● *What are the important times of the day?* (For example: meal times, school, bedtime and so on.) *Do these change?* (Consider holidays and weekends.)
● *What happens to the hour hand as we change the minute hand? Why is it moving?*

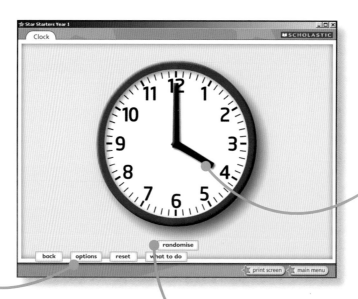

'options'
• Select setting for 'randomise' function
• Use to link clock hands

time set
Drag hands to change time

'randomise'
Press to select a random analogue time

Pictogram: favourite foods

Whiteboard tools
- Put the cursor in the white boxes to type in the title and label the categories of the pictogram.
- Press the arrow next to the red square to choose the symbol you want to use for the pictogram.
- Drag and drop the symbol chosen up to ten times to any row on the pictogram.
- Press 'reset' to start a new pictogram

What to do
Ask about half of the children in class to tell you their favourite foods and begin to build the pictogram. (Each block can only be dragged up to ten times to avoid the pictogram becoming too large.) Build the pictogram as the children are giving their favourites, or ask each child to come to the board to drag their own favourite to the correct position on the pictogram. Ask simple questions about the data.

Differentiation
Less confident: ask the children to count up the numbers representing each food item.
More confident: ask the children to tell you the favourite food item of all the children who were asked. How do they know this from the pictogram?

Key questions
- *How many children like bananas best?*
- *Which is the favourite food?*
- *How many more children prefer bananas to crisps?*

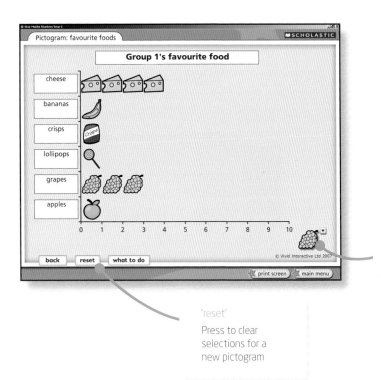

food item symbol
Drag and drop each symbol to build the pictogram

'reset'
Press to clear selections for a new pictogram

Block graph: favourite pets

Strand

Handling data

Learning objective

Answer a question by recording information; present outcomes using block graphs

Type of starter

Refine

Whiteboard tools

● Press on the white text boxes to type in the title and label the categories of the graph.
● Press the blocks of the graph to colour them and record an entry.
● Press 'options' to extend or change the scale of the graph.
● Press 'reset' to start a new graph.

What to do

Prepare a chart entitled 'Favourite pets' and type a selection of popular household pets in the white text boxes. Ask the children to tell you their favourite pets. (It is unlikely to be needed, but the vertical scale on the graph can be extended up to 20 using the 'options' button.) Build the block graph as children vote for their favourites. Ask the children to hold up their hands for each category or to come to the board to place a block on the graph (you may need to support them with this). Ask simple questions about the data.

Differentiation

Less confident: show the children the number on the vertical scale, representing each pet, or ask them to count the number of coloured blocks to answer questions such as: *How many children like dogs best?* Provide the children with coloured blocks as additional support.

More confident: ask the children to tell you the favourite pet of the whole class. How do they know this from the block graph?

Key questions

● *What is the favourite pet? How do you know?*
● *How many children like dogs best?*
● *Which animal had one more vote than the rabbit?*

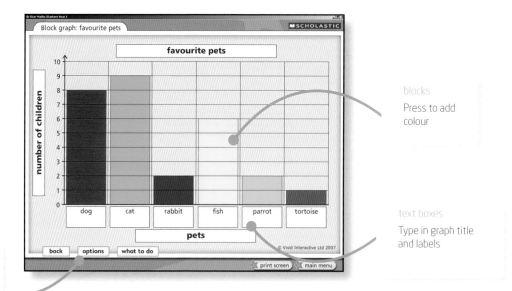

'options'
Adjust the scale of the graph

blocks
Press to add colour

text boxes
Type in graph title and labels

Targets

■ Find the target numbers using the cards below.

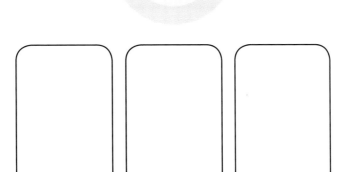

■ How I worked out the target answer.

_____

_____

_____

_____

_____

_____

_____

_____

SCHOLASTIC
www.scholastic.co.uk

Star Maths Starters ★ Year 1
PHOTOCOPIABLE

Beanstalk

Maps and directions

Plan your route using the map below.

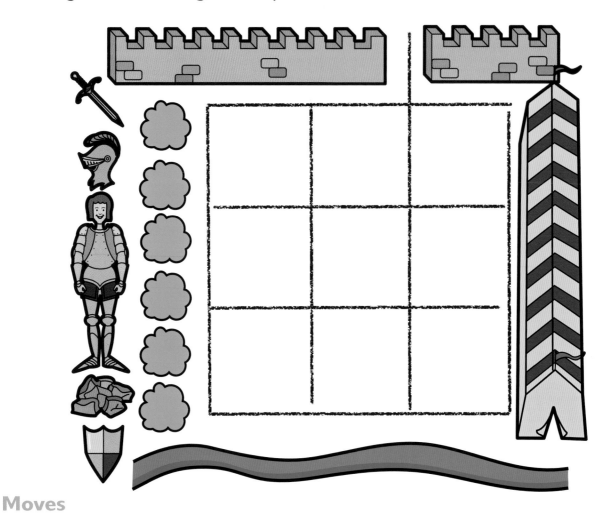

Moves

_____

_____

_____

_____

_____

_____

_____

Star Maths Starters diary page

Name of Star Starter	PNS objectives covered	How was activity used	Date activity was used

SCHOLASTIC

Also available in this series:

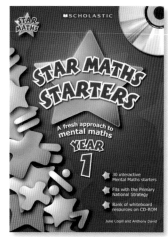

ISBN 978-1407-10007-4

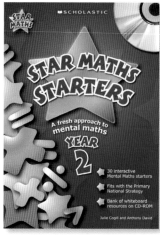

ISBN 978-1407-10008-1

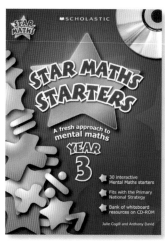

ISBN 978-1407-10009-8

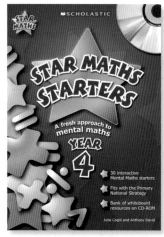

ISBN 978-1407-10010-4

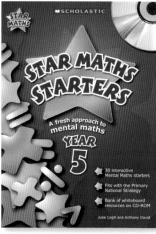

ISBN 978-1407-10011-1

ISBN 978-1407-10012-8

ISBN 978-1407-10031-9

ISBN 978-1407-10032-6

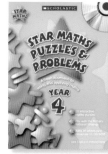

ISBN 978-1407-10033-3

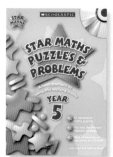

ISBN 978-1407-10034-0

ISBN 978-1407-10035-7

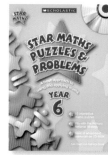

ISBN 978-1407-10036-4

To find out more, call: 0845 603 9091
or visit our website www.scholastic.co.uk